GOD

Does Not Play Dice

GOD
Does Not Play Dice

TOM JAMES

Library of Congress Control Number:		2014920152
ISBN:	Hardcover	978-1-5035-1488-1
	Softcover	978-1-5035-1489-8
	eBook	978-1-5035-1490-4

This book was printed in the United States of America.

Rev. date: 03/17/2015

To order additional copies of this book, contact:
Xlibris
1-888-795-4274
www.Xlibris.com
Orders@Xlibris.com
688970

CONTENTS

PREFACE

It has been said that one of the most complex challenges facing physicists today is coming up with a unifying theory that fully explains the origins of our universe and how the fundamental forces driving it relate to one another and to matter.

One day while surfing the net for information on a unifying theory I stumbled upon a video clip of Albert Einstein and had an amazing insight. At first this insight seemed small and obvious yet when I put this minor epiphany together along with Einstein's efforts at forming a unifying theory my eyes must have bugged out. This one little thing suddenly blossomed into complete clarity in that it might allow me to succeed in the exact line of research that Einstein believed to be the best path at finding a unifying theory!

While the journey I went on from there was no walk in the park, filled with both failures and successes, after eight thousand hours of work, I believe that what you are about to read may be the greatest discovery in physics since relativity, something that highlights a critical mistake physicists have made, the key to an effortless and easy understanding of our universe and the very first physics book ever published that explains "everything" using known proven physical laws only.

ACKNOWLEDGEMENTS

First and foremost I would like to acknowledge Albert Einstein, whose shoulders I'm standing upon, and Isaac Newton for pointing the way. Without the lifetime efforts of these two men, this book could never exist.

Next I would like to thank Stephen Hawking for his books and all the science shows and magazines available in Canada for providing enough general knowledge to give me a rich awareness of our universe over the years.

A special thanks to Danish Khan for his generosity in explaining the aspects of electromagnetism in planets and to the Central Connecticut State University for publishing the initial version of this Unified Field Theory in their *International Journal of Multidisciplinary Thought.*

I would also like to acknowledge and thank NLP Canada and Landmark Worldwide for providing me with extraordinary tools to go way beyond what I thought was truly possible.

To Rochelle Rodney, Susan Wilburn, Judith Snow, Connie Kostiuk, Dale Landry and Tallie Rabin I want to add a special thanks for their generous friendship that each gave at critical times in my life.

To Trini, my wife, all I can say is that I've been blessed to be living with someone who has a beauty that runs so deep she must be from heaven. And if it were not for my children, Kristina and Michael, who often challenge me, surprise me, and inspire me in ways I don't think they can fully appreciate I may not have been so aware of the gift that life is and that every day is a miracle.

To my brother, Gordon, I want to thank you for being the best friend a person can hope for, someone I can always rely upon to be real and straight even when I don't like it.

Lastly, I want to thank God for the gift of life and the amazingly wondrous gift that life truly is.

INTRODUCTION

When I was young, my parents would often take my brother and me camping during the summer months. One of my fondest memories of that time was when we camped at Meldrum Point on Manitoulin Island. The people that owned the land lived in a lighthouse that overlooked Lake Huron and told strange stories of ghost ships and UFOs. They said that the water surrounding the point was hazardous and that there was a strong undertow. Lastly, they told us that there was a magnetic reef in the water and that it made the air surrounding the point clearer than that in most places.

At night you could see the effect of the reef on the environment by the way it parted the clouds in the sky like some big invisible knife. I remember staring at the sky for four hours straight almost moved to tears in total awe of the panorama playing before my eyes. In the hyperclear air, I saw many more stars visible than I had ever experienced before. It was then that the universe captured my heart and my imagination. It reminded me of the cinematic power of the initial scene in *Star Wars*, which was so different from any other movie of its time and displayed stars in all their wonder.

Much later in life, when I began to take a serious look at physics, the universe began to look stranger and stranger. As I became aware of one illusion after another, it seemed like reality itself had taken a wrong turn somewhere and that some grand magician was at work when it came to the universe. Sensory deception, quantum weirdness, and even the nature of language itself all seemed to contribute to hiding the design of our universe. It is no wonder that even though many attempts have been made to pierce this sophisticated veil of illusion that humankind's best and brightest have been fooled. Einstein himself could not explain "everything," and although this was the case, he strongly believed that there was a distinct and definite design to the universe. To this end, Einstein is quoted as saying, "God does not play dice with the universe."

Four years ago, I had what I believed to be a profound insight regarding the work of Albert Einstein.

Before this insight, I had thought that a complete and unifying theory of how the universe worked was way too complex for me to comprehend and that even if I was lucky enough to come up with such a theory, there was no chance in hell that I could figure out the math that went along with it.

After having that insight, everything changed. I was so inspired that I thought I had a good chance of coming up with the theoretical part. And even though this endeavor turned out to be a lot tougher than I even remotely imagined after endless hours of work and ten thousand calculations, it all came together quite suddenly. It surprised and pleased me, and I believe I not only got the theoretical part to "everything" but that the math as well.

The insight I had that so inspired me was that I saw that a Unified Theory of Relativity could be formed with astonishing ease and that it would be only one step away from what Einstein had wished to accomplish by uniting General Relativity and Electromagnetism.

In brief, what I believe I have actually accomplished is that I have expanded upon Albert Einstein's most credible theories to form a unifying theory of the universe that explains "everything." By standing on the shoulders of giants like Einstein, Hubble, Hawking, and Newton, I may have completed what Einstein started. In the end, the new theory looked like some bizarre integration of Relativity, Classical Mechanics, and a variation of the Big Crunch Big Bang theory.

It has been said that Einstein wanted to see into the mind of God. Had I met Einstein, I would have said to him that before seeing the mind of God, we would need to first pierce the veil of illusions that stand in the way.

IN THROUGH THE LOOKING GLASS

A painter should begin
every canvas with a wash of
black, because all things in
nature are dark except where
exposed by the light.
                    —Leonardo da Vinci

A few years ago, the company I worked
for had a party for its high-end clients. They
had great food and drink, hired someone
who could roll genuine Cuban cigars and
a magician who went around from group
to group demonstrating his craft. When
the magician came to our group, he asked

me to pick a card, memorize it, and show it to the person next to me. He then asked me to place it back in the deck, which he shuffled. He then pulled out a card and asked, "Is this your card?" It wasn't. He shuffled the cards again and tried again with the same result. At that very moment while I was thinking that they had hired a lousy magician, he asked me to cut the cards. Once I did this, he asked me to turn over the top card. It was my card.

The hand is quicker than the eye, and while magicians practice the art of deception, few take the time to examine the machinery that gets deceived. We take our lives for granted, never quite waking up to the fact that the world is full of illusions. There is a great quote from a movie called *Joe Versus the Volcano* that reminds me of this, which states,

"My father says that almost the whole world is asleep. Everybody you know. Everybody you see. Everybody you talk to. He says that only a few people are awake and they live in a state of constant total amazement." For me, this is the world of physics.

Most people live life with their minds fully engaged and distracted by things of importance to them, not really aware that our minds and senses are being fooled all the time. For example, to a five-year-old, the moon and the sun appear about the same size, the sun appears to be travelling across the sky, and on a busy street, it appears that traffic is moving and that buildings are still. All these are illusions: the first due to a difference in distance, the second because of how our view changes as the Earth spins, and the third in the way that gravity affects objects sitting on a spinning object such as Earth.

Fig. 1 Distance Illusion

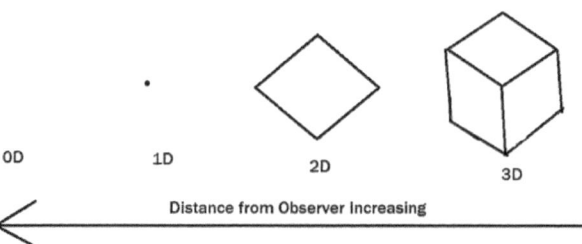

OD   1D   2D   3D

Distance from Observer Increasing

The farther something is away from us the smaller it "appears" to be. Althought the sun is over 400 times larger than the moon since it is also over 400 times the distance from us than the moon both the sun and the moon look about the same size when we look into the sky. We also lose our perception of the dimension of an object the further it is away. All stars except our own sun are so far away that most appear as one dimensional points that radiate light.

Fig. 2 Position Illusion

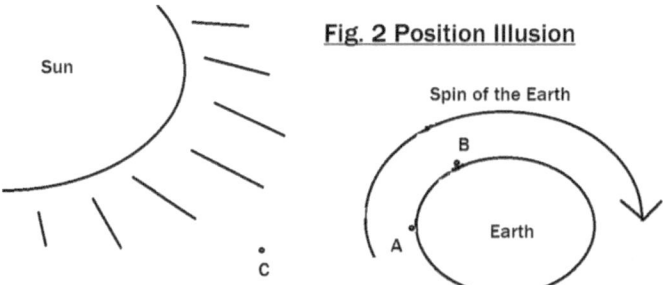

The Observer at 'A' sees the sun come up over the horizon at 'B' and as the Earth spins the sun "appears" to travel across the sky until it is directly overhead. That the sun is travelling around the Earth can be seen clearly as an illusion if you observe from point 'C.' In a sense the sun is not travelling across the sky and it is actually a case where the sky is travelling across the sun.

Fig. 3 Gravity Illusion

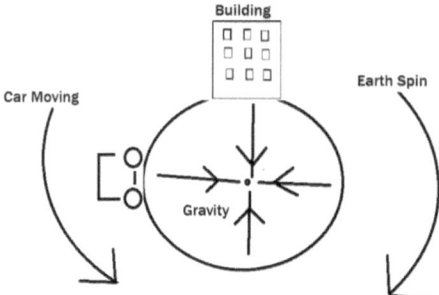

Gravity or the push to the center of the Earth holds both the car and the building to the surface. Although it "appears" to those on the surface that the car is moving and the building is still since the Earth is spinning both are actually moving.

While most adults can understand these illusions enough to explain them to a child, there are things that adults take for granted or accept as real that are nothing but distortions of reality. Imagine for a moment that you walked downtown in New York City and actually saw a group of naked cavemen huddled around the fire in the street, a bunch of North American Indians riding on horseback, a *Tyrannosaurus Rex* feeding on another dinosaur while all surrounding this weird scene were people walking, drinking coffee, and casually talking on their cell phones. If this was what you actually saw before you, I'd bet the last thing on your mind would be that this was "real." At best, you probably would have thought that you had stumbled onto some movie set or that someone had spiked your drink with a hallucinogenic. Yet, as any physicist will tell you, when you look at the stars in the night sky such a wacky distorted view is exactly what you are seeing. What I am referring to is Albert Einstein's discovery of the finite or fixed speed of light and how it distorts our view of the universe.

Light travels at 299,792,458 meters per second, yet our universe is so vast that even this high rate of speed is just not enough to "see" things as they are in the present. The light from our sun takes about eight minutes to get to Earth, so the sun, in its totality, all the light we see, is a view of events that happened eight minutes ago. The light from a star a billion light-years away, what we actually "see" is actually a view of that star from a billion years ago. In other words, when we look out at the night sky, what we are actually "seeing" is a collage of past views.

Fig. 4 Finite Lightspeed Illusion

Three stars in the current night sky

Star A

Four Light
Years Away

Star B

One Hundred Light
Years Away

Star C

One Billion Light
Years Away

As the distances between most objects in the universe is so vast one of the standard measures used is called the "light year" or the dustance that light travels in a year. Since light has been shown to have a finite or fixed speed one of the consequences of this is that when we look at the night sky it may "appear" that we are seeing what's happening right now yet this is an illusion as the light we are seeing is a collage of events that happened at many different times. In the above example what we see happened four years, one hundred years and a billion years ago. It may even be that the star we are seeing doesn't even exist any more.

Welcome to the twilight zone, the world of the physicist, where the web of illusions is so thick that I believe that sometimes, being their own worst enemies, even the physicists themselves get fooled. The one illusion that I believe most physicists have been blind to is the belief that our universe arose out of "nothing."

To be specific, it is the contention of the Big Bang Theory that our universe was once in a state where "nothing" existed, no matter, no energy, no space, and even no time and that our universe first appeared and spontaneously arose out of such a state.

For me, the baffling part about this contention was that when I searched for some evidence, any evidence at all, to support this belief of a universe arising from nothing, I found a mountain of facts and information that seemed to discredit any such belief. What I found instead was that a universe arising from "nothing" would violate the Law of Conservation of Energy, that a Big Bang or explosive expansion of our universe from "nothing" had no explanation whatsoever with regard to cause and effect and that such an occurrence seemed to contradict all observable evidence I could find in similar events such as Novas and Supernovas, the data involving the evolution of stars, and even the fact that matter always, always, always seems to be present even in our earliest views of the universe.

Either our universe arose from "nothing" or it arose from "everything." The latter view is what is known as the Big Crunch Big Bang Theory, and one of the contentions or implications of this theory is that matter was present before, during, and after the

Big Bang. Both of these theories about how our universe began are almost unthinkable in terms of human logic. Either you believe our universe really did arise out of nothing like some outrageously impossible magic trick, or you believe our universe never had a starting point ever and is merely engaged in some impossible endless cycle. However, if you consider that matter was present "always," something very remarkable begins to emerge.

Occam's razor states that the simplest explanation with the fewest number of assumptions is likely to be true. For example, it's a rainy night and you see a flash through your window. Is someone taking pictures of your house, or is it just lightning? Simply put, if there is no matter present at creation, forming a unifying theory is so complex no physicist has ever come up with one, whereas with enough matter present at creation, this single fact actually leads to and unifies every other aspect of how the universe works, fits perfectly with what we can observe, does not violate any laws, and the known attributes of the Big Bang are predictable in terms of cause and effect.

The attributes of the Big Bang are a singularity, insane heat, and the rapid expansion of the universe. In a universe arising from nothing, there really are no "cause-effect" facts that explain those attributes from nothing.

Edwin Hubble deduced that in the distant past, at the time of creation, all matter must have been in the same geographic location. With enough matter in the same geographic location, General Relativity states that at some point the gravity would be so incredibly strong that nothing could stop it from collapsing to a singularity. Of course, what would happen next would be an explosion of insane heat sending any residue matter and everything else out in every direction.

Fig. 5 Occam's Illusion

Occam's Razor : The theory with the simplest explanation and the least amount of assumptions is likely to be true.
Occam's Judgement : The theory with the greatest amount of assumptions is likely to be more complex, false and an illusion.

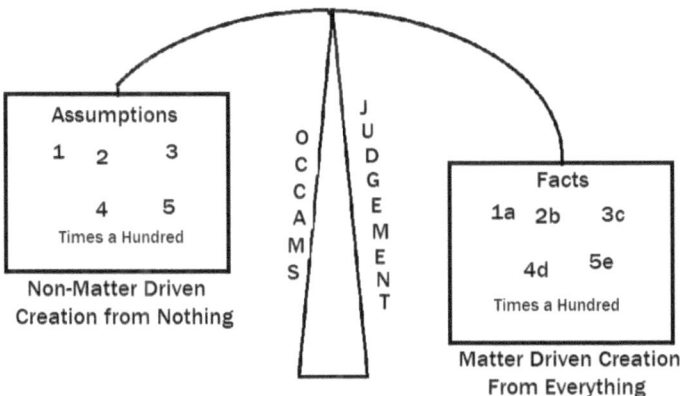

Assumptions 1 2 3 4 5 Times a Hundred **Non-Matter Driven Creation from Nothing**	**Facts** 1a 2b 3c 4d 5e Times a Hundred **Matter Driven Creation From Everything**

1 M-Theory	1a Hubble's Big Crunch
2 Singularity from Nothing	2b General Relativity Singularity
3 Insane Heat from Nothing	3c Big Crunch Big Bang
4 Too Hot for Matter to Exist	4d Evolution of Stars & Law of Conservation of Energy
5 Unexplained Acceleration Dark Matter Dark Energy	5e Newton's Equation f = ma

If you want to begin to see what others have missed or been in the dark about, allow yourself to be open to the idea of a matter-driven creation. For example, with matter present at creation, it immediately solves a problem physicists have been struggling with for decades. The entire physical universe is accelerating or moving faster than known sources can account for. Physicists have speculated or theorized that the cause of this additional acceleration was Dark Matter or even Dark Energy.

However, with matter present at creation, the acceleration of our universe is easily explainable by Classical Mechanics alone with a little formula by Isaac Newton $f = ma$, where f is force, m is mass, and a is acceleration!

This is the very beginning of a theory that explains everything that other theories can't or that they have difficulty in explaining. There is another obvious fact that supports the proposition of a matter-driven Big Bang.

On the surface this fact may seem to be insignificant, one that few physicists are likely to attach any great importance to yet, when rigorously examined, may be more critical in our understanding of the universe than anything else. The one observable fact that is actually quite odd is that "all matter is moving in the universe."

Fig. 6 Non-Motion is an Illusion

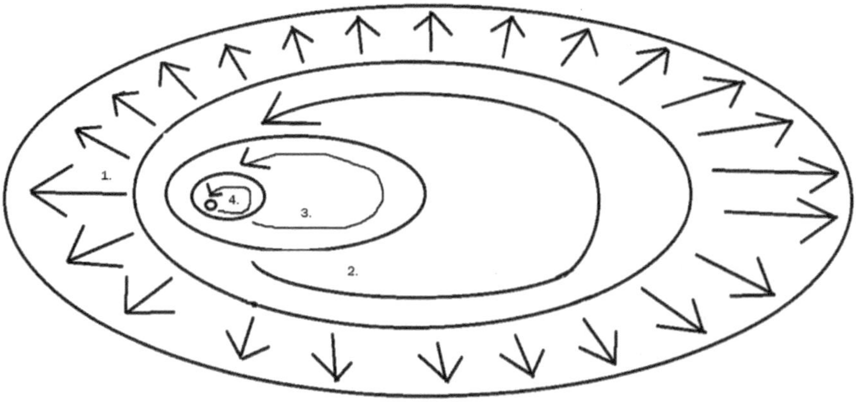

1. Entire Physical Universe expands or moves outward
2. Milky Way Galaxy moves around the Universe
3. Our Solar System moves around the Milky Way Galaxy
4. Earth moves around our Solar System

SHAKE, RATTLE AND ROLL

Give me a place to stand
and with a lever I will
move the whole world.
— Archimedes

Sometimes, even noticing the smallest things can have profound implications in science. When Alexander Fleming noticed mold growing in a petri dish, it led to the discovery of penicillin and the saving of countless millions of lives. In physics, the fact that "everything is moving" falls into such a category.

Earth is heavy. The sun is heavier still, and the solar system and galaxies even heavier. And when you look out into the universe at any object whatsoever, you will find that all objects are moving no matter how much resistance to motion they have due to mass. Again, I want to remind you that the entire physical universe is moving or accelerating much faster than sources can account for.

While there are a number of possible sources that can cause acceleration, if you consider force caused this motion, due to the extreme mass of some of the objects in our universe, pretty much the only force strong enough to cause this effect across the entire physical universe is the source at the beginning, an explosion, or the Big Bang itself.

A second odd thing about the observable motion in our universe is "how objects are moving." The movement observed universally is <u>only</u> consistent with two forces causing motion. Given that there were two forces causing motion and that Edwin Hubble has shown that our universe is expanding uniformly outward, what this suggests is that there were two explosive events or Big Bangs, one stronger than the other, in opposition to each other, driving all matter in our universe.

With these two early forces and matter present as described, there are five possible resultant motions:

(1) The motion given by an Expanding Universe.

(2) The second is a Net Direction through space like when you hit a baseball north with a westerly wind present.

(3) The third is Spin like when you hit a baseball north when there is a sudden gust of wind from another direction.

(4) The fourth type of motion comes from the interaction of multiple large objects in close proximity to one another where the previous motions and gravitational differences combine to give Orbital motion.

(5) The fifth type of motion is Gravity itself.

Note that all these types of motion are observable on the macroscopic level in our universe and that the discovery that gravity is a form of motion is the heart of something profound as to how the universe really works.

With matter present in the beginning, the principles and laws of motion are far more important with regard to the design of our universe especially where these proven known aspects of motion show up in an observational view of reality.

THE SUPERSTRUCTURE
OF OUR UNIVERSE

"Yes, that's it!" said the Hatter
with a sigh. "It's always tea time . . .
Well, I never heard it before, but
it sounds uncommon nonsense."
—Lewis Carroll, *Alice in Wonderland*

When I was young, I had an overly inquisitive nature, which often got me in to all sorts of trouble, such as when I drank perfume and had my stomach pumped, when I took apart our television to fix it, or when I got stuck in the heating ducts under the floor. If I had been born today, someone would have probably diagnosed me with Attention Deficit Disorder and put me on Ritalin. As it was, however, it led me to asking all sorts of questions, which didn't exactly make me popular in school. Sometimes the teacher didn't know the answer, or the answer they gave wasn't the answer I was seeking.

One area I had such unfulfilled questions around was the topic of gravity: "What is gravity?" "How does it work?" "Can you make it?" While most of the answers centered on the "effect" of gravity, such as how it bends space or that it attracts, they did not address what "caused" or created gravity in the first place.

What I am about to reveal is an alternative explanation for how gravity occurs, and while it may be difficult for some to believe, if you put aside your skepticism and just follow along by the end of this section, you will understand how gravity is a form of motion, how gravity is created, and how gravity makes up a kind of universal superstructure for matter to exist in.

Isaac Newton noticed that there was an attraction between large objects in space and postulated that the force causing this attraction was an attribute of matter itself, which he called gravity.

Albert Einstein asserted that mass was the attribute of matter that caused gravity and that the attractive effect was due to the extensive mass of large objects and the capacity of that mass to bend the space surrounding that object.

Gravity is not an attribute of matter itself, and while this notion may seem radical, the key to seeing this is, ironically, a discovery made by Albert Einstein.

In *Time Magazine*'s journal on the life of Albert Einstein, it mentions that Einstein noticed that there was no difference between a man standing on an object that was being accelerated upward and that same man standing on Earth. In other words, there is a type of equivalence between gravity and acceleration. The journal goes on to say that this eventually led Einstein to develop his General Theory of Relativity.

My point regarding Einstein's discovery is that the reason that there is equivalence between gravity and acceleration is that they are <u>exactly</u> the same thing!

Before I outline for you how gravity is just a form of motion we call acceleration, I want you to imagine for a moment that an object sitting on the floor before you moved suddenly. If this object were not an animal or a machine, the last thing you would assume was that the cause of the motion was a property of the object itself. Most likely you would conclude that you observed a type of acceleration or the effect of a force acting on an object!

In order to show you how gravity is a type of acceleration and why it appears to be an attribute of matter, I am now going to present to you a little thought experiment that involves a more rigorous look at Newton's Laws of Motion, relates to the beginning of the universe with matter present, and that defines the formation and structure of gravity in our universe.

Imagine you are in the frictionlessness of space in a space suit in a region of space far away from Earth and that it is completely empty of all matter except for you, a box surrounding you that is a million miles in length, width, and height and where in the center of this box is an object the size and mass of Earth. Next, imagine there was a tremendous explosion in close proximity to the object strong enough that it propelled that object at a hundred thousand miles an hour away from you.

As an observer, what you would be witnessing is Newton's Laws of Motion on a macroscopic scale involving a single force acting on an object, causing motion.

Stepping outside the thought experiment for a moment, I will now briefly explain the Laws of Motion at play thus far. Please note that some of the phrasing will be a bit more rigorously defined than is normally used and that there will be some additions as well.

Newton's First Law of Motion states that an object at rest stays at rest and an object in motion stays in motion. Both of these conditions stay that way unless acted upon by another (external) force. Also, the greater the mass an object has the greater its resistance is to motion.

Since all matter in the universe is moving all the time, nonmotion is an illusion.

A corollary to the aspect of resistance is that once an object is in motion, the greater the mass an object has, the greater its resistance is to changes in motion.

This is known as inertia, and the greater the mass an object has, the longer the accelerated portion of the original force will be preserved over time in the same direction as the force that acted upon that object.

Since a moving object adopts the attributes of the force that caused it to move, has the capability to impact other objects with the stored force given by inertia, and since all matter is moving all matter is by definition a "force."

Also, any object acted upon by a force that exceeds its resistance <u>must</u> move.

An object caused to move and acted upon by a single force will have <u>one</u> resultant motion along <u>the path of least resistance</u> directly away from the source of that force. Every atom in that object will be moving "uniformly" with the same speed and in the same direction.

Newton's Legacy : Part 1 The Foundation - A Single Force Acting Upon an Object
Fig. 7a Newton's Equation F = MA (Principle View)

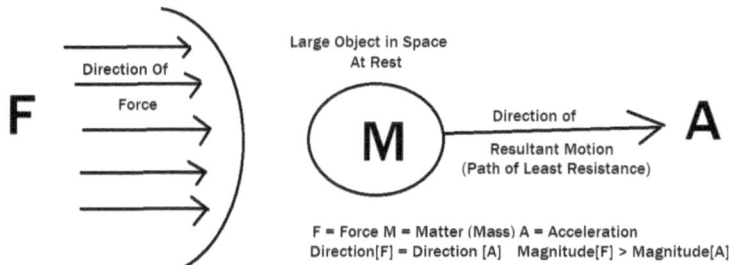

F = Force M = Matter (Mass) A = Acceleration
Direction[F] = Direction [A]   Magnitude[F] > Magnitude[A]

An overwhelmingly powerful force like that which results from a Nova or Supernova acts upon a large object in space causing it to move. Note that in this case the force greatly exceeds the objects resistance to motion so when the force hits the objects surface it causes a "uniform" motion directly away from the source of the force, the path of least resistance. One Force = One Resultant Motion

Newton's Legacy : Part 1 The Foundation - A Single Force Acting Upon an Object
Fig. 7 b Newton's Equation F = MA(Reality View)

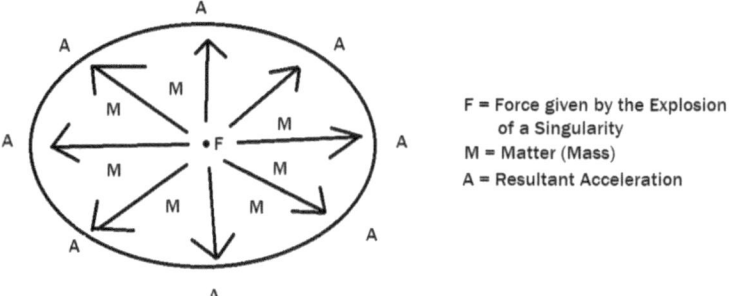

F = Force given by the Explosion
of a Singularity
M = Matter (Mass)
A = Resultant Acceleration

Like Fig. 7a having only one force acting on objects gives only one resultant motion. In this case the motion called "Expansion".

Since, in reality, more than one motion is observable on a large scale in our universe there must be more than one force acting upon all objects. While an obvious choice is gravity for this second force an equally obvious choice is a second explosion from a singularity or a second Big Bang.

Returning to the thought experiment, you watch the planet-sized object approach the edge of the box away from you when all of a sudden the entire edge explodes with a force ten times the intensity of the earlier one and hits the object slightly off center.

At that moment, you notice several things all happening at the same time. The first is that you notice that the object behaves like it hit an immovable brick wall, and due to the power of this new overwhelming force, you observe the object reversing its course, hurtling back toward you at a rate many times faster than it had been moving away. The second thing you notice is that since the force hit slightly off center, the object is now spinning and travelling at a slight arc toward you. The third thing you notice is that the object appears to have shrunk slightly in size and that it has become spherical in shape. The last thing you notice is that you have begun to move and that as the object moves toward you, your speed toward it seems to be increasing.

Fig. 8a Expansion (Principle View)

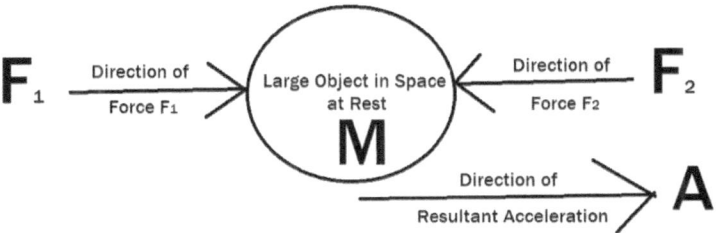

F$_1$ and F$_2$ = Forces where F$_1$ = 10 F$_2$, M = Mass, A = Acceleration

Two forces that greatly exceed an objects resistance to motion cause an object to move which, in this case, is the same direction as force F$_1$ due to it's stronger force.

Fig. 8b Expansion (Reality View)

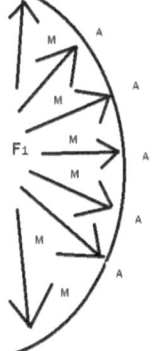

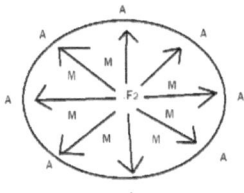

F$_1$ and F$_2$ = Force given by the Explosion of a Singularity where F$_1$ = 10F$_2$

M = Matter (Mass)

A = Resultant Acceleration

While the net result is a motion of "Expansion" as in the single force model as the force F$_1$ will overwhelm the force given by F$_2$ due to the interaction of conflicting forces and directions through matter other motions will also show up in addition to expansion.

Fig. 9 Net Direction (Both Principle and Reality View)

North

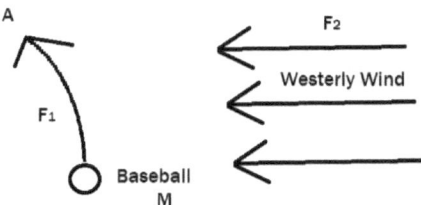

A baseball is hit north with a westerly wind blowing. The resultant Net Direction motion of the ball is somewhat northwest along a curve.

F₁ and F₂ are forces, M = Matter (Mass), A = Acceleration

As you can see from Fig. 8b when the forces hit some will hit at an angle as shown above and the Net Direction motion will be another type of motion visible to us in reality.

Fig. 10a Spin (Principle View)

North

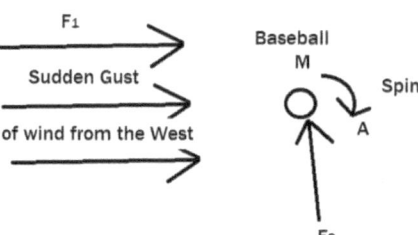

F₁ and F₂ are forces, M = Matter (Mass), A = Acceleration

A baseball is hit north when a sudden short gust of wind hits the ball causing it to spin.

Fig. 10b Spin (Reality View)

F_1 and F_2 = Force given by the
Explosion of a Singularity where $F_1 = 10F_2$
M = Matter (Mass)
A = Resultant Acceleration

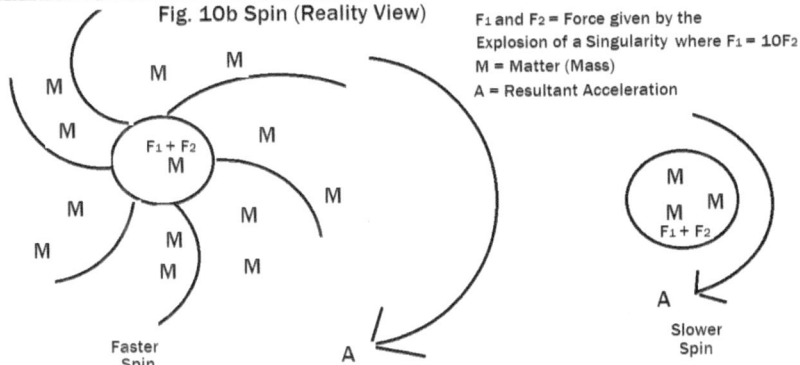

In reality when the two forces hit as shown in Fig. 8b it is very much like throwing two rocks in a pond, in the case of the universe where one rock is much bigger than the other. Each rock generates ripples going out from the center yet when they cross the net result is a motion of Spin as shown above. Initially what you see in the water is something that looks exactly like the above left or many of the spiral galaxies we can see in our universe. When the motion slows it loses its arms and you get a spinning circle or, in our universe a perfect sphere as is seen in most large objects.

Fig. 11 Orbital Motion (Both Principle and Reality View)

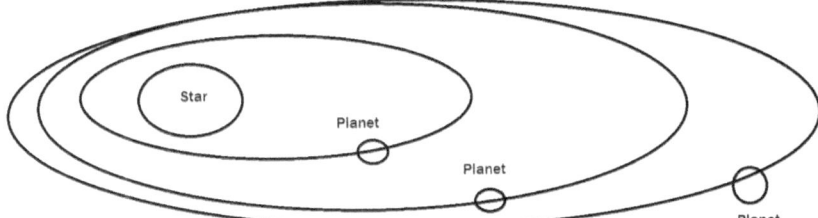

In Fig. 8b you can see if two Big Bangs interacted, one stronger than the other, when these powerful forces hit each other there would be absolute chaos, collisions, insane heat, matter shooting off in all directions and all the prior motions in action. As the universe expanded due to the variances in gravity the chaotic dance would form into a more ordered type of motion such as the one above called Orbital motion . This order would be moons around planets, planets around stars, solar systems around galaxies and galaxies around the universe.

While this is an imperfect thought experiment as you were not affected by the first force, there is enough information to know how the Laws of Motion relate to the second half of the experiment culminating in the emergence of the force we call gravity.

As stated earlier, if you consider that matter was present at creation and also that there were two matter-driven explosions or Big Bangs, then the effect on large objects with a lot of mass is very similar to the events of thought experiment.

Each of the two Big Bangs is strong enough alone to exceed any object's resistance to motion so that <u>any</u> object present at creation <u>must</u> move.

What happens is that the objects caught in between these two powerful forces are "crushed" to a certain degree. In terms of the resultant part of the motion causing the crush, every atom will "uniformly" move in a direction away from both of these powerful forces with the same speed. Since these two forces are in opposition to each other there is only one direction these atoms can move in.

In other words, caught between two powerful forces, all atoms crushed in the object <u>must</u> move "uniformly" along <u>the path of least resistance</u> toward the very center of the object.

Here this force of motion driving atoms to the center encounters another problem. If you travel in a car at a high rate of speed into a brick wall and you have no seat belt, you are going to have a bad day. What you notice in such a case is that quite suddenly you and all the force governing motion "move" to where the wall is. And so since the interaction between these two powerful forces causes the atoms to move to the center, an impenetrable wall, all the force moves to the center from every direction. Here something very special happens that is exactly like a magnifying glass.

Put a magnifying glass in the path of the sun, and soon you get a visible beam that is very hot and very bright. What happened is that many individual beams of light were all bent toward a common focal point where the heat and brightness from many beams land in the same geographic location, giving rise to the powerful amplified effect we see.

And this is exactly what happens with large objects caught between two overpowering forces. Matter acts as a lens, which focuses and bends the force of acceleration toward its center powerfully, where this attribute or "uniform push to the center" is amplified so much it extends beyond the circumference of the object into space.

The last piece of the puzzle is truly magical and is governed by the science involving inertia. Large objects with a lot of mass have a lot of inertia and have a great tendency to preserve forces over a long period of time, so just as large objects continue to spin, move through space, and orbit one another, this effect of great force moving toward to the center is preserved as a "structure" that persists over time.

Newton's Legacy : Part 3 Blueprints for Gravity
Fig. 12 The Crush

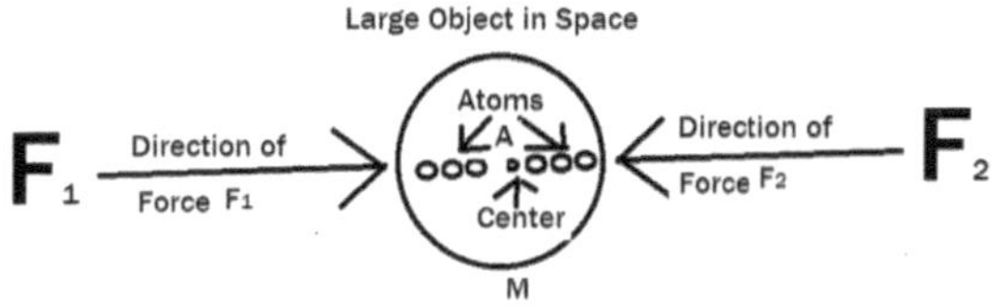

Large Object in Space

$$F_1 \xrightarrow{\text{Direction of}}$$
Force F₁

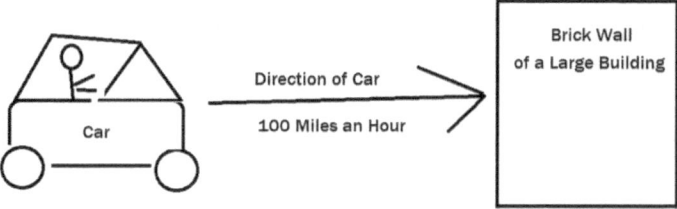

Atoms
A
000 ₀000
Center

M

Direction of
Force F₂ $\xleftarrow{\hspace{2cm}} F_2$

F₁ and F₂ = Force given by the Explosion of a Singularity where F₁ = 10F₂

M = Matter (Mass), A = Resultant Acceleration

In Fig. 8b you can see that some large objects will be caught exactly between these two powerful forces or Big Bangs being "crushed." While this effect actually occurs in all objects as the forces merge together into the form of an expanding universe the above highlights the effect of powerful opposing forces in a large object. Both forces greatly exceed the objects resistance to motion so that all atoms in the object <u>must</u> move "uniformly" along the path of least resistance which is toward the center of the object. All atoms are structurally connected powerfully to all other atoms so the effect is "uniform" movement from the surface of the object to the center, a crushing effect as the forces bend each other in that direction.

Newton's Legacy : Part 3 Blueprints for Gravity
Fig. 13 The Wall

Brick Wall
of a Large Building

Direction of Car

100 Miles an Hour

Car

A man without a seat belt travels in a car at 100 miles an hour toward the brick wall of a large building. When the car hits the wall the man flies out through the windshield toward "where the wall is" along with "all the force" causing motion.

Newton's Legacy : Part 3 Blueprints for Gravity
Fig. 14 Optics

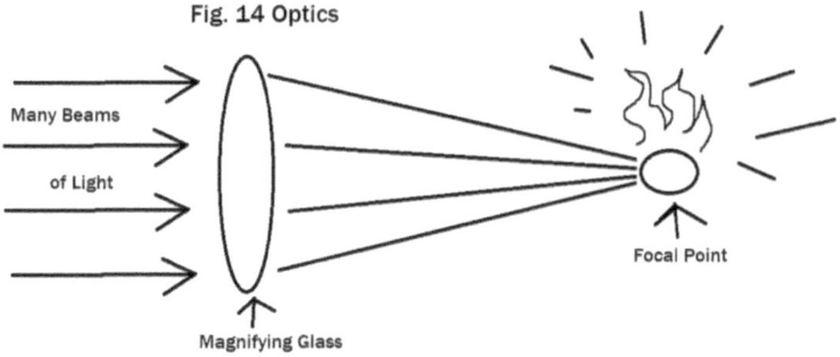

Many Beams

of Light

Focal Point

Magnifying Glass

Every beam of light from the sun has a little heat and a little brightness. The magnifying glass bends many beams of light all toward a common focal point that amplifies those two attributes of light by stacking them in the same geograpic location giving a very visable effect that is very hot and very bright.

Newton's Legacy : Part 3 Blueprints for Gravity
Fig. 15 Inertia

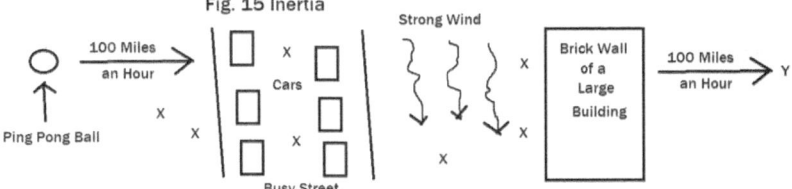

A ping pong ball is sent at a 100 miles an hour toward a large building.
X = Some of the potential end locations of a ping pong ball weighing an ounce.
Y = The result of a ping pong ball weighing 100 tons sent at a 100 miles an hour.

In space, once an object is in motion, the greater the mass an object has the greater its resistance is to changes in motion and the longer the force causing that motion will persist over time in both magnitude and direction. This is called inertia.

Newton's Legacy : Part 3 Blueprints for Gravity
Fig. 16 The Sructure of Gravity

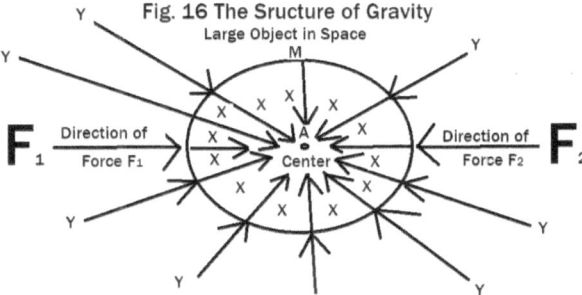

F_1 and F_2 = Force given by the Explosion of a Singularity, M = Matter (Mass), A = Resultant Acceleration, X = Fig. 12 The Crush, y = Fig. 14 Optics, The Center = Fig. 13 The Wall

Our Universe begins in dualistic explosions of force, heat and matter. The magnitude of the event and each of those three aspects is so great as to defy human comprehension and yet the above process outlines how gravity was redefined by that event upon all matter where inertia transforms forces causing motion into a "structure" of force causing gravity. Gravity, by it's nature, provides links between large objects in our universe, a Superstructure that persists over time that has danced and evolved through interaction from the beginning giving us moons around planets, planets and their satellites around stars, solar systems around galaxies and galaxies around our universe.

In the beginning there were two explosions, one stronger than the other. The effect on the objects with the greatest mass was a preservation of a percentage of those initial powerful forces over time as a type of Gravitational Superstructure, which imposes its structural will on all objects in the universe. In retrospect, it's a bit obvious. The moon goes around the Earth. The Earth and all the other planets go around the sun. Our solar system and all others in our galaxy travel around the Milky Way. Our galaxy and all others travel around the universe even as the entire physical universe spins and expands outward.

Up until this point, we have covered the beginning of our universe related to large objects, motion, and the gravitational structure resulting from the two initial

Big Bangs. Turning to the microscopic world, we now look at the structure caused in small objects caught between these overwhelming forces.

In the beginning there so much matter present it is comparable to having all matter that currently exists being present in the same geographic location in two large fields. While the effect of the forces on large objects gave us the large-scale motion we actually see in the universe and the gravitational structure that accompanies it, on extremely small objects, due to their lack of resistance to motion, the energy of those two forces drove particles together, causing fusion and the formation of atomic structure as opposed to gravitational structure.

A curious oddity of something in common between large objects and small that also supports the idea of two Big Bangs is the curious fact that both atoms and the dominant large objects in space are all "universally" spherical in shape.

Believe it or not, we have now covered three of the four fundamental forces required for a unifying theory and shown how they drive our universe. The Gravitational Superstructure, its persistence over time, and its influence on matter across space keeps fusion and the formation of atomic structure as a permanent aspect in existence.

The two Big Bangs in the unifying theory are known as the Strong Force and Weak Force, and we have talked a lot about the third, which is Gravity. The only force left to talk about is Electromagnetism.

Objects in space with electromagnetic fields are basically big electromagnets, and the only difference is that gravity affects the strength of the field by altering the path of free electrons. To have an electromagnet, you must have an object that contains electromagnetic material, a source of free electrons, and a spin either of the object or of the electrons around a still object. An object in space that has these three aspects will have an electromagnetic field.

There you have it, a unifying theory that covers how the fundamental forces relate to one another and matter. In the next section, I will cover some other aspects of interest and the math.

EUREKA

One of the greatest discoveries
a man makes, one of his great
surprises, is to find he can do what
he was afraid he couldn't do.
                              —Henry Ford

To form a unifying theory of the universe or a theory that unites "everything," it is said that this theory should show the relationship between the fundamental forces that drive our universe and matter. In other words, this theory, if possible, must show the relationship between the <u>Strong Force</u>, the <u>Weak Force</u>, <u>Gravity</u>, <u>Electromagnetism</u>, and <u>Matter</u>.

One way of knowing if such a unifying theory or Unified Field Theory is possible is if each of the underlined aspects above can be put into a single equation. Without defining specifics, you can tell if a series of fields can be put into the same equation if each field shares something in common with every other field and has some form of mathematical relationship or equivalence. After four years of research, I have found that there are three things in common among all these fields: Each has a relationship to matter, each is a form of energy, and each is a force.

To explain and clarify this a bit further, both the Strong and Weak forces have to do with what holds atoms or matter together, are obviously forces by name, and involve how energy is being held captive in atoms.

Gravity has a relationship to matter through mass, is a force that pulls, and has a form of force/energy that extends into space.

Without certain types of matter with certain properties, the aspect we call Electromagnetism could not exist; it is a force that influences certain types of matter, and similar to gravity, it has a force/energy that extends into space.

As stated earlier, matter is, by definition, a force. Under Special Relativity, matter has a relationship to energy.

Since each of these aspects of a unifying theory have three things in common with every other aspect then it can be concluded that these aspects can all be put into a single equation, which means that a unifying theory of the universe is, in fact, possible.

The ideal equation then to begin to develop a Unified Field Theory is the one that has each of the three common aspects or force, energy, and a relationship to matter. One equation that meets this criteria is the one that gave us the force of the atom bomb or Einstein's equation $e = mc^2$.

Ironically, this is the same equation I had such a profound insight of around four years ago. Albert Einstein once said in reference to this equation, "A small amount of mass can be converted to a large amount of energy and vice versa." In the vice versa case, I saw something interesting with respect to General Relativity.

General Relativity states that there is a direct relationship between mass and gravity, where the gravitational strength is determined by the amount of mass. One implication of this direct relationship is that mass and gravity are inseparable phenomena and that neither can exist without the other being present. In the vice versa case above, or $m = e \div c^2$, at the moment mass exists, gravity must also be present, yet this was clearly missing in the equation!

Fortunately, there is an equation that exists called the Gravitational Constant, which shows the exact mathematical equivalence between mass and gravity such that gravity can easily be incorporated into Einstein's equation. By doing so, it could be said that this new equation would correlate to a Unified Theory of Relativity.

What sparked my imagination about this was that Einstein had thought that if he could unite General Relativity and Electromagnetism, he could gain enough insight into how "everything" works that he could develop a unifying theory of the universe, and at least mathematically, this new equation was only one step away from what Einstein dreamt of doing.

To me, seeing this, it felt like Christmas morning with a stocking full of presents and nothing to do but to play all the time. While getting Electromagnetism into the equation turned out to be a lot harder than I anticipated, being this close to accomplishing something Einstein had failed to do spurred me on over the next four years, and in the end, I succeeded.

Overwhelm: Have you ever had to cram for an exam and found that you had too much to remember, or started a new job where there seemed too much to do, or even been confronted by all the curveballs life sometimes throws you? This is exactly how I felt when I began to study Electromagnetism.

Initially, I thought it was going to be simple, that planets were just big electromagnets, and that as long as they were spinning, had free electrons, and enough electromagnetic material, I could easily get the information I needed.

My hypothesis was that there must be a relationship between the amount of electromagnetic material and the strength of the electromagnetic field generated, that the amount of electromagnetic material present could be expressed as a percentage of the overall mass, and that I could use that to plug into the Einstein's expanded equation.

Using Earth as a standard for electromagnetic field strength through which the fields of other planets in our solar system would be measured against, I first got the rate for Earth or how much electromagnetic material generated how much of its electromagnetic field.

When I then tried to apply this Earth rate to other planets in our solar system as a predictor of their electromagnetic field strength, they all failed. As far as I could tell, my hypothesis was in error.

It was then I met a physicist named Danish Khan, who specialized in Electromagnetism. He said it was not as simple as I made it out to be, that there were three types of electromagnetic material, that Earth had two electromagnetic fields, and that there were a lot more things to take into account in determining the electromagnetic field of a planet, including the fact that some planets didn't have them. I felt as if I had fallen into some deep hole.

And the more I researched, the deeper the hole seemed to get. I looked at Maxwell's equations, how the electromagnetic field worked on Earth as electrons passed through the planet, and briefly considered Quantum Gravity as a way of finding equivalence between gravity and electromagnetism.

And then out of the abyss, a thought struck me or, rather, a thought experiment regarding my original hypothesis. If the rate changed with more electromagnetic material, this might be the problem skewing my results.

The problem with testing the math was that I also found it difficult to get any reliable data for the three types of electromagnetic material for other planets, let alone Earth. Even focusing on "iron" alone was a challenge, as the latest data from satellites seemed to conflict with known book data. Then a thought experiment I invented resolved this problem, which I called the Blind Dart Player.

Imagine you are blindfolded, given some darts, and asked to hit the bull's-eye of a dartboard in front of you. With enough darts, you will eventually hit the bull's-eye.

Even though I did not have an accurate figure for the amount of electromagnetic material each planet had, I knew that it had to have been between 1 and 100 percent of the overall mass and I could pick all the percentages to see what comes up.

After tens of thousands of calculations, I was able to get correct predictions of electromagnetic field strength relative to any two planets, yet it utterly failed again in finding any stable predictor constant relative to electromagnetic material alone.

Again, it seemed my initial hypothesis was in error.

Then something occurred to me that the calculations were pointing to.

Back in the thought experiment, if someone were moving the dartboard back and forth like a pendulum whenever you hit the bull's-eye, it would likely be in a different location. What all the calculations pointed to, as they were not off by that much, was that there was an additional element at play that was skewing the results.

And then the obvious hit me as a possibility. What if the total mass itself and the corresponding gravity were somehow skewing the results, affecting how strong an electromagnetic field you got by its effect on electrons. And when I plugged this factor into the more than ten thousand equations, everything worked, and all the field strengths of the planets in our solar system with electromagnetic fields were predicted exactly.

In fact, when I simplified the equation, something bizarre occurred in that the equation resolved into an Electromagnetic Constant relative to mass!

In the end, it turned out that my original hypothesis was correct with a slight alteration. As long as an object in space has free electrons, is spinning, and has electromagnetic material, the electromagnetic field strength can be predicted by the amount of electromagnetic material it has and by the amount of mass it has in totality.

Now, before I give the resulting equation for you to look at, for a brief moment I want to say something about Special Relativity and how it applies to the modified equation I've developed for a unifying theory. In the two cases Einstein referred to in his quote, mass to energy and energy to mass, it should be noted that they involve opposite types of reactions, fission and fusion.

While most people are familiar with the mushroom cloud that results from the detonation of an atomic bomb, in space such a detonation, without Earth being in the way, would appear more like a fireworks display or an explosion of energy radiating outward spherically, starting from a central point.

Now, if we were looking at a movie of such an event and ran the film backward, what you will then see is the opposite effect or the fusion of energy into matter (mass). In other words, what you will see is an energy sphere that shrinks before your eyes to a central point where matter (mass) is then formed.

What is interesting about this reverse movie is that this fusion reaction involves pure energy manifesting into matter (mass), and the "nature of how it behaves" or how the energy acts in terms of motion is identical to how matter behaves in the presence of two powerful opposing forces forming gravity. It's as if fusion of energy into matter (mass) or an atomic structure and the formation of a gravitational structure both share a common "cause."

This duality of design drives fusion, gravity, and the Superstructure that persists over time. This duality is reflected in the Special Relativity equation, where the numerator denominator aspects reflect the fusion of any two energetic forces or any two elements of matter into other matter regardless of magnitude including the Strong and Weak forces that created our universe.

It took four years to get the following Unified Field Equation, and while it is by no means perfect or complete, further research, testing, and development involving this equation may reveal whether or not our universe will collapse again to a singularity, open up alternative ways to create artificial gravity, and reveal new resources and methods for attaining clean energy.

The Unified Field Equation and its development are listed below:

Special Relativity—Fission
$$e = mc^2$$

Special Relativity—Fusion
$$m = e \div c^2$$

Inclusion of Gravity (Raw)

$$m + g = e \div c^2$$

Inclusion of Gravity Mass Equivalence (Standard Gravity) and Quantity Variables for Different Magnitudes of Force/Mass:

$$q1m + (((G \times q1m) \div r^2) \div 9.822)\, g = q2e \div q3c^2$$

Inclusion of Electromagnetism (Raw) and a Presence Variable

$$q1m + (((G \times q1m) \div r^2) \div 9.822)\, g + nEM = q2e \div q3c^2$$

Final Unified Field Equation

$$\mathbf{q1m + (((G \times q1m) \div r^2) \div 9.822)\, g + n\,((m \times f) \div K)\, EM = q2e \div q3c^2}$$

e = energy, m = mass in kilograms, c = speed of light, g = gravity

q1, q2, q3 = quantity, G = Gravitational Constant = 6.67×10^{-11}

r = radius, 9.822 meters per second squared = Earth's surface gravity

f = amount of iron mass in kilograms a planet or sun or object has

K = Electromagnetic Constant = 12.474315×10^{48}

EM = electromagnetism, n = 0 or 1, depending whether or not the planet or star in question has all the elements necessary to generate an electromagnetic field (i.e., Free electron source, iron or other electromagnetic material, gravity, and the aspect of spinning).

THE TOUGHEST QUESTION OF ALL

It's like the man said. Never
take life too seriously. You
will never get out of it alive.
— Bugs Bunny

While my equation is not perfect, I
believe it is perfect enough and that the
theory is sound enough to safely say that
our universe has an order and organization
to it, a design by any other name.

And if our universe has a design, we
are then forced to confront the ultimate
question regarding the existence of our
universe, "Is the design of our universe

an intentional act, or is it a random occurrence, or is it merely the product of known physical laws?"

Those that hold that this design was an intentional act of God in the creation of the universe point to something called the Anthropic Principle or that the universe was designed for human life. Further, they point to something called Fine Tuning, the idea that the many constants in our physical universe together provide the ideal conditions for life where, if you change any of these constants in any way whatsoever, results in no life in our universe. Examples of this are the size of atoms, how fast the universe is expanding, the fact that stars explode providing the non-Earth elements for life to occur.

While it is debated how many of these constants are critical to the design of the universe and life, it is asserted that the many constants that relate to or contribute to life showing up together are too complex a series to be something that can ever be a mere random coincidence.

In response to this, physicists have posed something called M-Theory, or the theory of multiple universes. In their eyes, universes are being created all the time and that the reason the universe is the way it is with life is that we just happened to win the universal lottery. The problem with M-Theory is that it is purely theoretical and that there is absolutely no evidence for it whatsoever that there exists another universe other than our own.

Part of the assertion that our universe was a random creation is something called the Uncertainty Principle. In 1927, Werner Heisenberg said, "The more precisely the position is determined, the less precisely the momentum is known in this instant, and vice versa." This has to do with Quantum Mechanics and the idea that things that occur in the microscopic world are only predictable in terms of probability and that in the world of the very small, it is all essentially a random game. As the early universe occurred in a singularity, this would point to it being a randomized event.

There is a distinct problem, however, with this assumption, which I call the Principle of Relative Uncertainty. The problem is where Heisenberg's uncertainty meets Newton's Laws of Motion. The more mass an object has, once that object is in motion, the greater its resistance will be to changes in motion, and the longer force causing that motion will persist over time without change, preserving both the magnitude and direction. As such, the greater the mass an object has, the more predictable outcomes due to motion will be. In other words, the greater the mass, the greater the ability to predict the momentum and position of a given object and vice versa. This makes the Uncertainty Principle a predictable phenomenon at the quantum level.

However, if you include that matter was present in the beginning, the enormous amount required to cause two singularities, two explosive events, due to the mass involved under the Principle of Relative Uncertainty, it makes the outcomes due to motion absolutely predictable, the Superstructure of gravity and the constants in our universe predictable. In other words, our universe could not be any other way and that the design is definitely not a random phenomenon.

While this does not "prove" God exists, it certainly points to the possibility that God exists, something I did not truly expect to find in my explorations in physics.

It did, however, lead me to speculate on two things: Would it be possible for human beings to create a universe one day, and if God exists, how could have God come to exist in the first place?

Humanity has come a long way since hiding in caves and the discovery of fire. Technology these days is advancing exponentially. We make artificial materials, tactfully dabble in genetic engineering, have travelled to space, are printing living material on printers so as to create new organs, and I have just learned that an invention I thought of to bend and focus gravity is technically feasible. If humanity survives its own cleverness, who's to say what we can be capable of in a thousand, a million, or even a billion years? Do we even know if we are the only ones in the universe developing such potential?

Whether or not you believe we will ever be able to be advanced enough to create a universe from scratch, what I do know is that if all we learn is just how to move enormous amounts of matter in space into the same geographic location, we can cause singularities to occur, resulting in events that essentially duplicate what I believe caused the creation of our universe, which includes the creation of life itself.

As to how God might have come to exist given that it may have been that a Big Crunch was coming and if there were living sentient beings who knew this ahead of time, if they were sufficiently advanced enough, they may have found a way to survive that occurrence. It may be that their consciousness only survived by being part of the new universe, where that consciousness is imbued into every part of the physical universe.

Both the Bible and physicists insist our universe arose from nothing, yet everything I've found points to a matter-driven creation. Since a matter-driven creation and the Principle of Relative Uncertainty point to the possible existence of God, if God indeed exists and our universe arose from nothing, where it also was a matter-driven creation, the only way for both to be equally true is if the universe came about exactly as it states in the Bible, or first there was "nothing" and then there was "everything."

Whether or not you believe in God, I do believe that if God were to see what is going on in our world today, he would have another commandment for us, "Love one another." As such, I am including a last thing, not about physics, but about saving this planet of ours.

THE HAND OF GOD

Neither death, nor life, nor angels,
nor principalities, nor powers,
nor things present, nor things to
come. Nor height, nor depth, nor
any other creature, shall be able to
separate us from the love of God.
—Romans 8:38

Once upon a time, there was an elderly couple who lived on the outskirts of a small city. The woman was kind and told stories to the children who passed by her front gate each day. Her husband loved her very much and tended a beautiful big garden in their backyard. She would often think of the first time he gave her a flower. It was a very long time ago, when people weren't as bold and up-front in relationships as they are now. He was shy and nervous. Finally, he asked, "Can I put a flower in your hair?" She smiled and nodded. He approached her carefully and placed a deep-red rose in her hair. He bent to kiss her, tripped, and brought them both tumbling down.

They both started laughing and, shortly after, committed themselves to each other for life. From that day on, he gave her a flower every morning.

Years passed, and the neighborhood became more and more crowded and troubled. Crime and gangland activity were on the rise. In spite of this, the couple shared their love with all. Children still came to hear the stories and see the beautiful flowers in the garden. Whenever there was bad news, the old woman would pray in good faith for a resolution. It seemed to her that God listened, as the source of the distress would always lessen and disappear. People thought that they must be blessed to escape all the violence around them.

A month later, her husband was mugged and killed while going home from the store. She had felt as though someone had cut out a piece of her soul. She had thought they would be together when one of them was about to die, and prayed for some sign that he was at peace.

Several months passed, and she began to accept her husband's death, but without a sign, it never felt quite right. As with many elderly people who have lost their spouse, her health began to be affected, and soon she was spending all her time alone. She was in great need and decided to pray the same prayer for her husband that she had all those months ago. The next morning, there was a knock at the door. A little girl was standing on her doorstep with a small bunch of flowers in her hand. She hugged the old woman, smiled, and said, "I picked these for you for all the wonderful stories you told me." The girl hesitated a moment and added with a laugh, "Can I put a flower in your hair?"

The ultimate gift of story is
twofold; that at least one soul
remains who can tell the story,
and that by recounting of
the tale, the greater forces of
love, mercy, generosity, and
strength are continuously called
into being in the world.
—Clarissa Pinkola Esstes, PhD

We have all been blessed to feel the falling of rain on a warm and nostalgic day, to hear the windy whisper of nature's soul in the cries of playful children, and to see the colorful panorama that has defined the birth of a new day. You have been given life and the ability to create new life. These things are, as they are, for a reason. It is only who we are for one another that gives meaning and purpose to our existence.

Knowledge is not power unless it is used. Love is not divine unless it is expressed. Life is not the purest joy imaginable unless we use every fiber of our being in its realization.

When I finished writing and editing this story, it was about three o'clock in the morning. Even though I was tired, I did one more thing. I took some potted African violets from our front room and placed it at my wife's bedside along with this story. And that is exactly what I request of you. Act. Give some flowers along with this story to at least five people you hold dear, and ask that they do the same. Leave five more copies anonymously for others to pick up in places where you don't normally go. If this goes out to ten people, and those ten give it to ten others, within ten repetitions, this story will have gone out to the whole world. Smile and give thanks. We can all truly bring beauty and love to the world. We can all make a difference.

For me and, hopefully, for you, this book has answered a lot of the questions as to how the universe works and, if not, at least something that's made a difference for you.

Thank you.

INDEX

A

Acceleration, 32, 36, 43–44, 56

Atoms, 47, 55–56, 62, 65, 82

Attraction, 42

B

Bible, 88

Big Bang, 17, 29, 33, 36–37, 42, 44, 46, 54, 56, 61–62

C

Constants, 82–83, 85

Gravity, 38, 41–44, 54, 63–64, 66–68, 77, 79–80

H

Hubble, Edwin, 29, 37

I

Illusions, 14–15, 18, 20–21, 23, 26, 46

Inclusion of Electromagnetism and a Presence Variable, 79

Inclusion of Gravity, 79

Inclusion of Gravity Mass Equivalence, 79

Inertia, 46–47, 57

L

Law of Conservation of Energy, 27

Laws of Motion, 44–45, 54, 84

Light, 24

M

Mass, 32, 42, 45–46, 57, 66–68, 70, 73, 75–77, 80, 84–85

Momentum, 84–85

Motion, 36–39, 42–47, 54–55, 60–61, 77, 84–85

M-Theory, 83

N

Net Direction, 37

Nothing, 26–29

O

Object, 42–47, 49–50, 54–55, 57, 60–61, 63

Orbital motion, 38

P

Q

R

S